4-5歲

英文篇

Daily Conversation

日常英語會話

園丁文化

Good Morning and Good Night
早安和晚安

做得好！ 不錯啊！ 仍需加油

Put a ✓ in the correct box. 請在正確的 □ 內加 ✓。

1.

□ Good morning.

□ Good night.

2.

□ Good morning.

□ Good night.

答案：1. Good morning. 2. Good night.

Good Afternoon and Good Evening
午安和晚安

做得好！ 不錯啊！ 仍需加油！

Fill in the blanks with the correct letters.
請在橫線上填寫代表正確答案的英文字母。

A. morning　　B. afternoon

C. evening　　D. night

1.

2.

12
1
2
3
4
5
6
7
8
9
10
11

Good ______, Dad !

答案：1.B 2.C

Good Night
晚安

做得好！  不錯啊！  仍需加油！ 

- What is the girl going to say? Write the correct answer on the line and circle the correct answer.
  小女孩想說什麼？請填寫和圈出正確的答案。

2. She is going to (school / bed).

答案：1. night　2. bed

Revision 1
複習（一）

Draw a line to match each picture with the correct greeting.
請用線把圖畫和正確的問候語連起來。

1. 

2. 

3. 

4. 

- Good morning.
- Good afternoon.
- Good evening.
- Good night.

答案：1. Good evening. 2. Good morning. 3. Good night. 4. Good afternoon.

Hello and Hi
你好

做得好！  不錯啊！  仍需加油！ 

- Help Peter visit each of his friends before he goes home.
  男孩想在回家前先探望朋友，請為他畫出正確的路線。

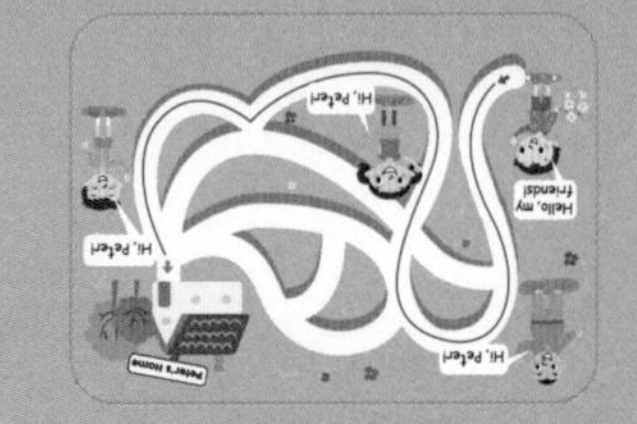

答案：

Goodbye
再見

做得好！  不錯啊！  仍需加油！  

Which pictures show people saying "Goodbye" ? Colour the correct 💬 .
哪些圖中的人物正在說 "Goodbye" ？ 請在正確的 💬 填上顏色。

1. 

Goodbye

2. 

Goodbye

3. 

Goodbye

4. 

Goodbye

答案：1、4

Hello and Goodbye
你好和再見

做得好！ 不錯啊！ 仍需加油！

- It's March. Winter has gone and spring has come. Fill in the blanks with "Hello" or "Goodbye".
三月，冬去春來。請在橫線上填寫 "Hello" 或 "Goodbye"。

Hello　　　**Goodbye**

1. ______________ winter !　　2. ______________ spring !

答案：1. Goodbye 2. Hello

Please
請

做得好！ 不錯啊！ 仍需加油！

Which of the boys is more polite? Colour his ☆ in red and trace the word "Please".
哪個男孩有禮貌？請在他旁邊的 ☆ 填上紅色，並沿着灰線寫一寫 "Please"。

答案：2. ★

Thank You
謝謝

做得好！ 不錯啊！ 仍需加油！

The girl is willing to lend pencils to the boys. Draw a line to connect the red and green pencils to the boys in the same colour top. What should they say in return? Put a ✓ in the correct box.
小女孩願意把鉛筆借給男孩們，請用線把紅色和綠色的鉛筆分別連至穿着相同顏色上衣的男孩。男孩們應該說什麼？請在正確的 ☐ 內加✓。

答案：Thank you!

I am Sorry and Excuse Me
對不起和不好意思

做得好！ 不錯啊！ 仍需加油！

What should the person say in each picture? Please colour the relevant word cards.
圖中的人物應該說什麼？請把相關的字卡填上顏色。

1.

I	night.	Good	
am	you.	Thank	sorry.

2.

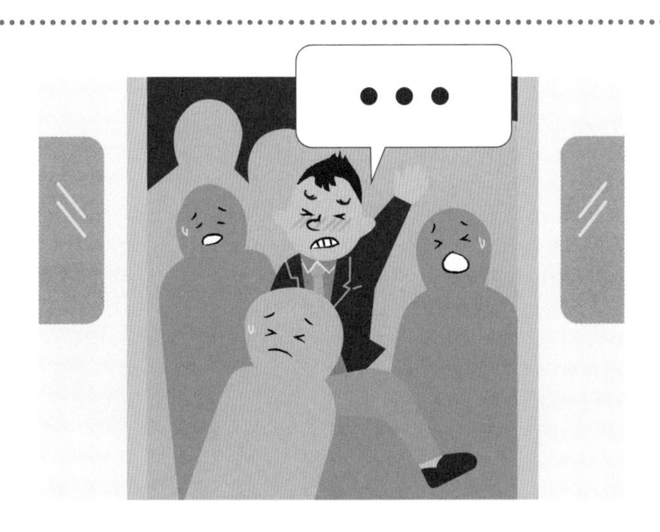

I	you.	Excuse	
am	sorry.	Thank	me.

答案：1. I am sorry. 2. Excuse me.

Revision 2
複習（二）

- Find and colour the following words. (The words could appear horizontally or vertically.)
請找出以下詞語，並填上顏色。（注意：詞語可以是橫行或直行顯示。）

HELLO HI GOODBYE
PLEASE THANK YOU

K	C	O	R	J	T	L	U
V	H	A	G	P	H	I	T
D	E	P	L	E	A	S	E
T	L	M	Q	I	N	Z	W
A	L	N	K	S	K	V	O
G	O	O	D	B	Y	E	B
Q	S	A	P	L	O	F	J
U	W	H	Y	T	U	H	N

答案：

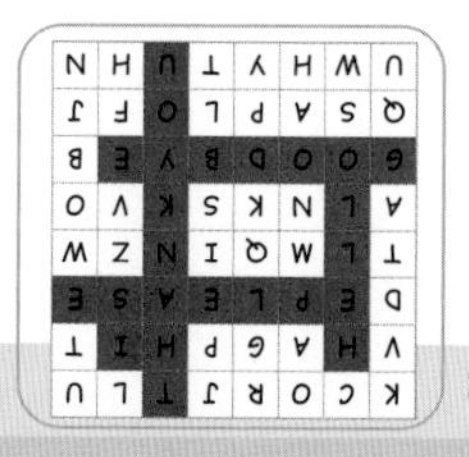

做得好！ 不錯啊！ 仍需加油！

Put a ✓ in the correct box. 請在正確的 ☐ 內加 ✓。

1.

☐ Thank you.

☐ Excuse me.

2.

☐ Thank you.

☐ I am sorry.

3.

☐ I am sorry.

☐ Excuse me.

4.

☐ Goodbye.

☐ Good night.

答案：1. Excuse me. 2. Thank you. 3. I am sorry. 4. Goodbye.

Happy Birthday
生日快樂

做得好！ 不錯啊！ 仍需加油！

- Fill in the blanks with the missing letters and draw the correct number of candles on the birthday cake.
  請在橫線上填寫正確的英文字母，並在生日蛋糕上繪畫正確數量的蠟燭。

答案：Happy Birthday

Merry Christmas
聖誕快樂

做得好！ 不錯啊！ 仍需加油！

Write "Merry Christmas" on the banner. Then circle five things that are not related to Christmas in the picture .
請在彩帶上寫 "Merry Christmas"，並在圖中圈出 5 件不屬於聖誕節的物件。

答案：

Happy Easter
復活節快樂

做得好！  不錯啊！  仍需加油！ 

- Colour the Easter eggs that spell "HAPPY EASTER".
  請找出能組成 "HAPPY EASTER" 的復活蛋，並填上你喜愛的顏色。

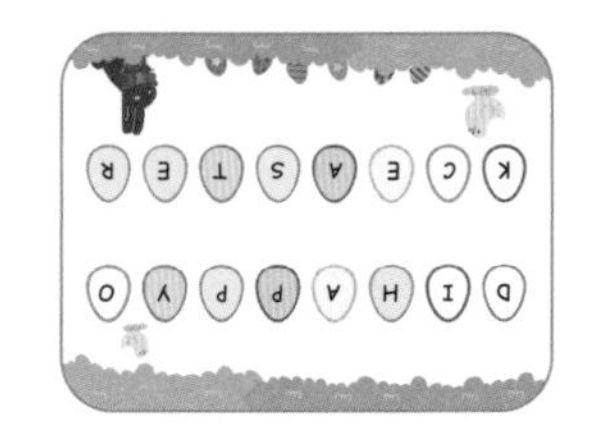
答案：

Revision 3
複習（三）

做得好！

不錯啊！

仍需加油！

Draw a line to match each greeting with the correct picture.
請用線把祝賀語和正確的圖畫連起來。

1. Happy birthday!

2. Happy Easter!

3. Merry Christmas!

A.

B.

C.

答案：1.B 2.C 3.A

Happy Mother's Day
母親節快樂

做得好！ 不錯啊！ 仍需加油

- Arrange the pictures in the correct order and write 1 to 4 in the correct box. 請按先後次序排列以下圖畫，並在 □ 內填寫 1 至 4。

A.

Daisy makes a necklace for her mum.

B.

Daisy's teacher talks about Mother's Day.

C.

Daisy gives the card and present to her mum.

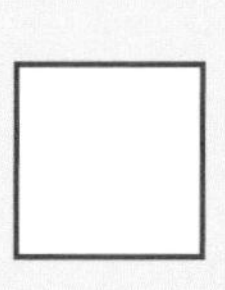

D.

Then Daisy makes a card for her mum. □

答案：A.2 B.1 C.4 D.3

Happy Father's Day
父親節快樂

- Read the card below, fill in the blank and circle the picture which shows Dad wearing the new tie. 請根據賀卡，在橫線上填寫正確的答案，並圈出爸爸結上新領帶的圖畫。

A. 

B. 

C. 

D. 

答案：Father, C

Get Well Soon
早日康復

做得好！  不錯啊！  仍需加油！ 

What could you bring when you visit a friend who is sick? Circle the things that are suitable. Then fill in the blanks with the missing letters to complete the sentence.
探望病人時，適合帶什麼物品？請圈出合適的物品，並在橫線上填寫正確的英文字母，完成句子。

1. 
Flowers

2. 

Snack

3. 
Fruit

4. 
Party popper

5. 

Get well card

6. 
Cake

Get w_ _ _ soon!

答案：1、3、5；well

Revision 4
複習（四）

做得好！ 不錯啊！ 仍需加油！

Draw a line to match each greeting card with the correct picture.
請用線把心意卡和正確的圖畫連起來。

1. 

•　　　　•
A. 

2. 

•　　　　•
B. 

3. 

•　　　　•
C. 

答案：1. C 2. A 3. B

I Love You and I Miss You
我愛你和我想念你

做得好！ 不錯啊！ 仍需加油！

Fill in the blanks with the missing letters to complete the sentences.
請在橫線上填寫正確的英文字母，完成句子。

1.

I l ___ ___ ___ you.

2.

I m ___ ___ ___ you.

答案：1. love 2. miss

Well Done and Try Again
做得好和請再試試看

做得好！ 不錯啊！  仍需加油！ 

Lead the boy to victory in the basketball match!
請畫出正確的路線，帶領男孩在籃球比賽中取得勝利！

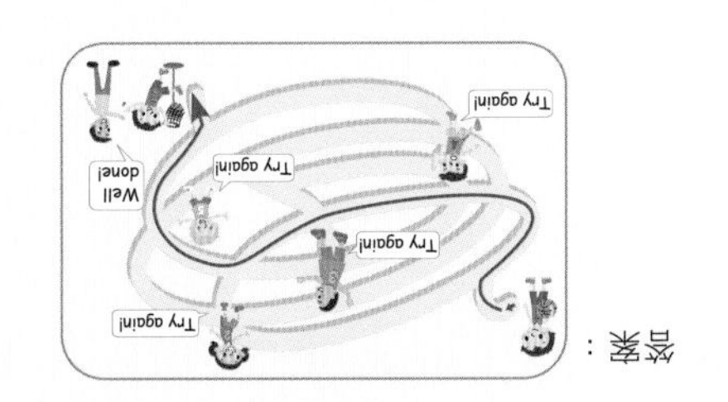

Stand Up and Sit Down
起立和坐下

做得好！ 不錯啊！ 仍需加油！

- Does the boy follow the teacher's instruction? Please colour the correct icons.
  小男孩有沒有聽從老師的指示？請把正確的圖案填上顏色。

1.

2.

答案：1. 2.

Introducing Myself (1)
介紹自己（一）

Introduce yourself by answering the questions below.
請回答以下問題，介紹自己。

1. I am ________ years old.

2. My name is ________________
____
________________.

3. 
I am a (boy / girl).

4. My favourite colour is ________________.

5. 
My hobbies are ________
____
________________.

6. My favourite food is ________
____
________________.

Introducing Myself (2)
介紹自己（二）

做得好！ 不錯啊！ 仍需加油！

Draw a line to match each description with the correct person.
請用線把描述文字和正確的人物連起來。

1. My name is Sally. I love cooking with my mum. My favourite food is pizza.

A. 

2. My name is Jack. I like playing football with my friends.

B.

3. My name is Anna. I want to be a doctor when I grow up. My favourite colour is blue.

C.

答案：1. C 2. A 3. B

My Family
我的家庭

- Fill in the blanks with the corret letters.
  請在 ⬭ 內填寫代表正確答案的英文字母。

A. Grandpa　　B. Grandma　　C. Dad

D. Mum　　E. Sister　　F. Brother

My Family Tree

答案：1.A 2.B 3.D 4.C 5.F 6.E

Introducing My Family
介紹我的家庭

做得好！ 不錯啊！ 仍需加油

- Draw a picture of your family and complete the sentences.
  請畫出你的家人，並完成句子。

My family

This is my family. I have ______ brother(s) and ______ sister(s). We live (on the Hong Kong Island / in Kowloon / in the New Territories).

Nice to Meet You
很高興認識你

Arrange the dialogue in the correct order and write 1 to 3 in the correct box.
請按先後次序排列以下的對話，並在 ☐ 內填寫 1 至 3。

☐ A. Hello Ken, I am Kelly.
Nice to meet you!

☐ B. Good morning, my name is Ken.
What is your name?

☐ C. Nice to meet you too, Kelly.

答案：A. 2 B. 1 C. 3

Ask and Answer Questions
基本問與答

做得好！  不錯啊！  仍需加油！ 

Read the dialogue carefully. Put a ✓ in the correct box and fill in the blanks with the correct words. 請仔細閱讀下面的對話，在正確的 ☐ 內加 ✓ ，並在橫線上填寫正確的答案。

1. ☐ A. How are you?
   ☐ B. What is your name?

   My name is Molly.

---

2. How old are you?

   I am ________ years old.

---

3. ☐ A. What do you like to do?
   ☐ B. What do you do?

   I like singing.

---

4. Nice to meet you!

   Nice to ______ you, too!

答案：1. B 2. four/ 4 3. A 4. meet

Asking Questions
發問問題

Choose the correct question for each occasion and write the correct letter in the ▭. 請為各個情景選擇正確的問題，並在 ▭ 內填寫代表正確答案的英文字母。

A. What do you do?

B. How are you?

C. What do you like to do?

D. What is your favourite colour?

答案：1. C 2. A 3. B 4. D

Revision 5
複習（五）

Look at the picture below and write the correct letter in each ▭.
請觀察下圖，在 ▭ 內填寫代表正確答案的英文字母。

A. Excuse me. B. Goodbye. C. Thank you.

D. Nice to meet you. E. Happy birthday!

答案：1. B 2. E 3. D 4. A 5. C